I0446341

THE FUNDAMENTALS OF CONSTRUCTION QUANTITY SURVEYING

Steven Smith, Ph.D.

Wisdom Publishers

ISBN: 9798864572580
Imprint: Independently published

Cover design by: Art Painter
Library of Congress Control Number: 2018675309
Printed in the United States of America

To all the quantity surveyors who make our world a better place, this book is for you.

You are the unsung heroes of the construction industry, working behind the scenes to ensure that projects are completed on time, within budget, and to the highest quality standards. You are the ones who measure and estimate the cost of construction work, manage contracts, and resolve disputes. You are the ones who use your skills and knowledge to make sure that our buildings, bridges, roads, and other infrastructure are safe and reliable.

This book is dedicated to you, in appreciation of your hard work and dedication. I hope that it will be a valuable resource for you as you continue to build your career and make a difference in the world.

Sincerely,
Steven Smith

The cost of a building is the price paid for it. The value of a building is the quality of the experience it provides.

NORMAN FOSTER

CONTENTS

INTRODUCTION

Quantity surveying is a complex and challenging profession, yet it remains essential for the successful completion of construction projects. Quantity surveyors play an important role in the construction industry, involved in all phases of a project, from the initial planning stage to the final completion.

The fundamentals of quantity surveying are crucial for quantity surveyors for some reasons. Firstly, it provides a foundation for grasping the construction process and its associated costs. Secondly, it prepares quantity surveyors to create accurate and realistic cost estimates, which are indispensable for budgeting and decision-making. Lastly, it furnishes the skills and knowledge necessary for effectively managing project costs throughout the construction process.

Understanding these fundamentals can assist quantity surveyors in preparing precise and realistic cost estimates for new construction projects, renovations, and repairs. Quantity surveyors are expected to leverage this knowledge to accurately estimate the quantities of materials and labor needed for a project, along with their associated costs. This helps to develop a comprehensive cost estimate for the project and, ultimately, to establish the project budget to ensure its financial feasibility. Once the budget is established, quantity surveyors proceed to create and maintain the bill of quantities, a detailed document listing all work to be performed on the construction project, along with quantities and costs.

Quantity surveying skills are essential for advising clients on the most cost-effective procurement strategy for their projects. This knowledge allows quantity surveyors to evaluate bids from contractors and make recommendations to the client. Quantity surveyors require these skills to negotiate contracts with contractors on behalf of the client and to help the client understand the contract's terms and conditions.

This proficiency enables efficient monitoring and control of project costs during the construction process. Quantity surveying knowledge prepares quantity surveyors with the ability to track actual costs against the budget and identify potential cost overruns. They can also collaborate with the contractor to identify and implement cost-saving measures.

Effectively managing variations and change orders becomes considerably easier with this knowledge. Variations and change orders involve changes to the scope of work on a construction project. Quantity surveyors can assess the cost and schedule implications of any variations or change orders and provide recommendations to the client.

These skills are also essential for preparing final accounts and settlements at the conclusion of a construction project. Quantity surveyors can calculate the final project cost and reconcile it with the budget. They can also prepare and issue payments to contractors and suppliers. A good understanding of the fundamentals of quantity surveying is essential for anyone who wants to succeed in the construction industry. This will enhance informed decision-making and promote success in project delivery.

CHAPTER 1: PRINCIPLES OF QUANTITY SURVEYING

1.1Measurement Units and Conversions

Measurement units are the standards by which we measure quantities. They are used in all aspects of our lives, from the simple task of measuring the height of a person to the complex task of measuring the distance to a star. In quantity surveying, it is important to be able to use and convert between different measurement units. This is because construction projects involve a wide range of materials and activities, each of which may be measured using different units. For example, a quantity surveyor may need to measure the following:

- The length of a piece of steel in meters
- The area of a floor in square meters
- The volume of concrete in cubic meters
- The weight of a load in kilograms
- The temperature of a room in degrees Celsius

Quantity surveyors also need to be able to convert between different units of measurement. For example, a quantity surveyor may need to convert from meters to feet when measuring the length of a piece of lumber. Some of the important measurement units and conversions in quantity surveying are:

- Length: Meters (m), feet (ft), and yards (yd)
- Area: Square meters (m²), square feet (ft²), and square yards (yd²)
- Volume: Cubic meters (m³), cubic feet (ft³), and cubic yards (yd³)
- Weight: Kilograms (kg) and pounds (lb)
- Temperature: Degrees Celsius (°C) and degrees Fahrenheit (°F)

It is important to be accurate when using and converting measurement units. In quantity surveying, even small inaccuracies can lead to significant cost overruns and delays. There are a number of ways to improve the accuracy of measurement units and conversions:

- Use accurate measuring tools and equipment.
- Follow standard measurement rules and practices.
- Double-check measurements.
- Have a good understanding of the construction process.

The following are some examples of how measurement units and conversions are used in quantity surveying:

Estimating the cost of a construction project: Quantity surveyors use measurement units to estimate the quantities of materials and labor required for a construction project. They then use this information to calculate the cost of the project.

Preparing the bill of quantities: The bill of quantities (BOQ) is a detailed document that lists all the work to be carried out on a construction project and the associated quantities and costs. Quantity surveyors use measurement units to prepare the BOQ.

Monitoring and controlling project costs: Quantity surveyors use measurement units to monitor and control project costs. They track actual costs against the budget and identify any potential cost overruns.

Managing variations and change orders: Variations and change orders are changes to the scope of work on a construction project. Quantity surveyors use measurement units to assess the cost and schedule implications of any variations or change orders.

- To convert from meters to feet, multiply by 3.28084.
- To convert from feet to meters, multiply by 0.3048.
- To convert from square meters to square feet, multiply by 10.76391.
- To convert from square feet to square meters, multiply by 0.09290304.
- To convert from cubic meters to cubic feet, multiply by 35.314667.
- To convert from cubic feet to cubic meters, multiply by 0.0283168.
- To convert from kilograms to pounds, multiply by 2.20462.
- To convert from pounds to kilograms, multiply by 0.453592.
- To convert from degrees Celsius to degrees Fahrenheit, multiply by 9/5 and add 32.
- To convert from degrees Fahrenheit to degrees Celsius, subtract 32 and multiply by 5/9.

Measurement units and conversions are an essential part of quantity surveying. By understanding and using measurement units accurately, quantity surveyors can help to ensure that construction projects are completed on time, within budget, and to a high standard of quality.

1.2 The Significance of Accurate Measurements

Accurate measurements are essential in all aspects of construction, but they are especially important in quantity surveying. Quantity surveyors use measurements to estimate the cost of materials and labor, prepare bills of quantities, monitor project costs, and manage variations and change orders.

Estimating the cost of a construction project

Quantity surveyors use measurements to estimate the cost of a construction project. They do this by calculating the quantities of materials and labor required for the project, and then multiplying these quantities by the unit costs of the materials and labor. If the measurements are inaccurate, the cost estimate will also be inaccurate. This can lead to cost overruns and delays. For example, if a quantity surveyor underestimates the amount of concrete required for a project, the contractor will need to order additional concrete. This will delay the project and increase the cost.

Preparing the bill of quantities

The BOQ is a detailed document that lists all the work to be carried out on a construction project and the associated quantities and costs. Quantity surveyors use measurements to prepare the BOQ. If the measurements in the BOQ are inaccurate, the costs in the BOQ will also be inaccurate. This can lead to disputes between the client and the contractor. For example, if a quantity surveyor underestimates the quantity of bricks required for a project, the contractor may charge the client for additional bricks that were not included in the BOQ.

Monitoring project costs

Quantity surveyors use measurements to monitor project costs. They track actual costs against the budget and identify any potential cost overruns. If the measurements are inaccurate, the quantity surveyor will not be able to accurately monitor project costs. This can lead to cost overruns. For example, if a quantity surveyor overestimates the quantity of concrete required for a project, the contractor will have leftover concrete. This will increase the cost of the project.

Managing variations and change orders

Variations and change orders are changes to the scope of work on a construction project. Quantity surveyors use measurements to assess the cost and schedule implications of any variations or change orders. If the measurements are inaccurate, the quantity surveyor will not be able to accurately assess the cost and schedule implications of variations or change orders. This can lead to disputes between the client and the contractor. For example, if a client changes the design of a window, the quantity surveyor will need to calculate the cost and schedule implications of the change. If the measurements are inaccurate, the quantity surveyor may not be able to accurately assess the cost and schedule implications of the change. This could lead to a dispute between the client and the contractor.

Accurate measurements are essential in quantity surveying. By using accurate measurements, quantity surveyors can help to ensure that construction projects are completed on time, within budget, and to a high standard of quality.

1.3 Standard Measurement Rules and Practices

Standard measurement rules and practices are essential for quantity surveyors to ensure that measurements are accurate and consistent. This is important for estimating the cost of construction projects, preparing bills of quantities, and monitoring project costs. Following standard measurement rules and practices can also help to avoid disputes between the client and the contractor, and improve the quality of construction projects. The following are some examples of standard measurement rules and practices:

- Measure from the outside of a wall to the outside of the opposite wall when measuring the width of a room. This is because the thickness of the walls is included in the overall width of the room.

- Measure from the finished floor level to the ceiling when

measuring the height of a room. This is because the height of the room is measured from the top of the finished floor to the bottom of the ceiling.

- Measure the longest dimension of a piece of material when measuring its length. This is because the length of a piece of material is determined by its longest dimension.

- Measure all materials and labor to the nearest whole unit. This is because it is not practical to measure materials and labor to a high degree of precision.

- Use consistent units of measurement throughout a project. This is important to avoid confusion and errors.

In addition to these general rules, there are also specific standard measurement rules and practices for different types of construction materials and labor. For example, there are specific rules for measuring concrete, bricks, lumber, and electrical work. Quantity surveyors can find standard measurement rules and practices in a variety of resources, such as building codes, industry standards, and quantity surveying textbooks and reference materials. These are some examples of how standard measurement rules and practices are used in quantity surveying:

Estimating the cost of construction projects: Quantity surveyors use standard measurement rules and practices to estimate the quantities of materials and labor required for a construction project. They then multiply these quantities by the unit costs of the materials and labor to estimate the total cost of the project.

Preparing bills of quantities: The bill of quantities is a detailed document that lists all the work to be carried out on a construction project and the associated quantities and costs. Quantity surveyors use standard measurement rules and practices to prepare the bill of quantities.

Monitoring project costs: Quantity surveyors use standard measurement rules and practices to monitor project costs. They track actual costs against the budget and identify any potential cost overruns.

1.4Precision in Quantity Surveying

Precision in quantity surveying is important for ensuring that cost estimates are accurate and that projects are completed on time and within budget. Precision refers to how close measurements are to each other, regardless of whether they are accurate or not. There are a number of ways to improve precision in quantity surveying, including:

- **Using accurate measuring tools and equipment**: It is important to use accurate measuring tools and equipment to ensure that measurements are as precise as possible.

- **Following standard measurement rules and practices**: Following standard measurement rules and practices helps to ensure that measurements are consistent and accurate.

- **Double-checking measurements**: It is always a good idea to double-check measurements to ensure that they are accurate.

- **Having a good understanding of the construction process**: Having a good understanding of the construction process can help quantity surveyors to identify potential areas of error and to take steps to minimize these errors.

The following are some examples of how precision is important in quantity surveying:

- **Estimating the cost of a construction project**: When estimating the cost of a construction project, quantity surveyors need to be able to accurately measure the quantities of materials and labor required. If the measurements are not precise, the cost estimate will not be accurate. This could lead to cost overruns or delays.

- Preparing the bill of quantities: The bill of quantities is a detailed document that lists all the work to be carried out on a construction project and the associated quantities and costs. If the measurements in the bill of quantities are not precise, the costs in the bill of quantities will also not be precise. This could lead to disputes between the client and the contractor.

- **Monitoring project costs**: Quantity surveyors use measurements to monitor project costs. They track actual costs against the budget and identify any potential cost overruns. If the measurements are not precise, the quantity surveyor will not be able to accurately monitor project costs. This could lead to cost overruns.

- Managing variations and change orders: Variations and change orders are changes to the scope of work on a construction project. Quantity surveyors use measurements to assess the cost and schedule implications of any variations or change orders. If the measurements are not precise, the quantity surveyor will not be able to accurately assess the cost and schedule implications of variations or change orders. This could lead to disputes between the client and the contractor.

CHAPTER 2: COST ESTIMATION

2.1The Art of Cost Estimation in Construction

Construction cost estimation is the art of predicting the costs involved in a construction project. It is a complex process that requires a deep understanding of the project scope, design, specifications, and construction methods. Estimators must also consider a wide range of factors, including material prices, labor rates, equipment costs, and overhead expenses.

While there is a science to cost estimation, there is also an art to it. Estimators must be able to draw on their experience and knowledge to make informed judgments about the costs of complex projects. They must also be able to anticipate and account for unexpected events. Accurate cost estimation is essential for the success of any construction project.

By accurately estimating the costs of a project, owners and contractors can:

- Set realistic budgets and avoid cost overruns
- Make sound decisions about the project scope and design
- Allocate resources efficiently
- Manage risks effectively
- Ensure that the project is completed on time and within budget

The art of cost estimation lies in the ability to make informed judgments about the costs of complex projects. Estimators must be able to consider a wide range of factors, including:

The project scope: This includes understanding the project's objectives, requirements, and constraints.

The design: The design of the project will have a significant impact on the cost of the project. Estimators must be able to understand the design and identify any potential cost drivers.

The specifications: The specifications of the project will also impact the cost of the project. Estimators must be able to understand the specifications and identify any potential cost drivers.

The construction methods: The construction methods that will be used to build the project will also impact the cost of the project. Estimators must be able to understand the construction methods and identify any potential cost drivers.

Material prices: Material prices can fluctuate significantly over time. Estimators must be able to obtain accurate quotes for materials and factor in the possibility of price fluctuations.

Labor rates: Labor rates can also vary depending on the location of the project and the type of labor required. Estimators must be able to obtain accurate quotes for labor and factor in the possibility of wage increases.

Equipment costs: Equipment costs can also vary depending on the type of equipment required and the length of time that the equipment will be needed. Estimators must be able to obtain accurate quotes for equipment and factor in the cost of transportation and setup.

Overhead expenses: Overhead expenses include costs such as insurance, permits, and office expenses. Estimators must be able to accurately estimate overhead expenses and factor them into the overall cost of the project.

Making Informed Judgments
Once the estimator has considered all of the relevant factors, they must make informed judgments about the costs of the project. This is where the art of cost estimation comes in. Estimators must draw on their experience and knowledge to make realistic and achievable cost estimates.

Anticipating and Accounting for Unexpected Events
Even the best cost estimators can't predict everything. Unexpected events can occur that can impact the cost of a project. For example, bad weather can delay construction and increase costs. Or, the discovery of unforeseen site conditions can require additional work. Estimators must be able to anticipate and account for unexpected events in their cost estimates. This can be done by including a contingency fund in the budget. The contingency fund should be used to cover the costs of unexpected events.

2.2 Preliminary Cost Estimation

Preliminary cost estimation is the process of estimating the cost of a construction project before the detailed design has been completed. Preliminary cost estimates are typically used to inform decision-making at the early stages of a project, such as whether to proceed with the project or to secure funding. Preliminary cost estimates are typically less accurate than detailed cost estimates, as they are based on less information. However, they can still be valuable tools for budgeting and planning purposes.

There are a number of different methods that can be used for preliminary cost estimation. Some of the most common methods include:

Order-of-magnitude estimate: An order-of-magnitude estimate is a very rough estimate of the cost of a project. It is typically based on historical data from similar projects.

Unit cost estimate: A unit cost estimate is based on the unit costs of different types of work, such as the cost per square meter of concrete or the cost per linear meter of electrical wiring.

Parametric estimate: A parametric estimate is based on a set of mathematical relationships between the cost of a project and certain project parameters, such as the size of the project or the complexity of the design.
The specific method that is used for a preliminary cost estimate will depend on the level of information that is available and the accuracy that is required.

Challenges of Preliminary Cost Estimation
There are a number of challenges associated with preliminary cost estimation, including:

Incomplete project scope: At the early stages of a project, the project scope may not be fully defined. This can make it difficult to accurately estimate the cost of the project.

Lack of detailed design: Without a detailed design, it is difficult to accurately estimate the quantities of materials and labor required to complete the project.

Uncertainty in material prices and labor rates: Material prices and labor rates can fluctuate over time. This can make it difficult to accurately estimate the cost of the project.

Despite the challenges, there are a number of things that can be done to improve the accuracy of preliminary cost estimates, including:

Use multiple methods: Using multiple methods to estimate the cost of a project can help to reduce the uncertainty in the estimate.

Use realistic assumptions: When making assumptions about the project scope, design, and construction methods, it is important to be realistic.

Include a contingency fund: A contingency fund should be included in the cost estimate to cover the costs of unexpected events.

Get feedback from experts: It is helpful to get feedback on the cost estimate from other experts, such as experienced cost estimators or contractors.

2.3 Detailed Cost Estimation

Detailed cost estimation is the process of estimating the cost of a construction project in detail. Detailed cost estimates are typically prepared once the detailed design of the project has been completed. They are used to budget for the project and to track costs throughout the construction process.

Detailed cost estimates are typically more accurate than preliminary cost estimates, as they are based on more information. However, they can also be more time-consuming and expensive to prepare.

There are a number of different methods that can be used for detailed cost estimation. Some of the most common methods include:

Quantity takeoff: Quantity takeoff (QTO) is the process of measuring the quantities of materials and labor required to complete a project. This is typically done using the construction drawings.

Assembly pricing: Assembly pricing is the process of pricing assemblies of materials and labor. Assemblies are groups of related items that are installed together, such as a wall or a roof.

Resource-based estimation: Resource-based estimation is the process of estimating the cost of each resource that will be required to complete the project, such as labor, materials, and equipment.

The specific method that is used for a detailed cost estimate will depend on the type of project and the level of accuracy that is required.

Steps in Detailed Cost Estimation
The following are the typical steps involved in detailed cost estimation:

Understand the project scope: The first step is to develop a clear understanding of the project scope. This includes understanding the project's objectives, requirements, and constraints.

Perform a quantity takeoff: Once the project scope is understood, the estimator must perform a quantity takeoff. This involves measuring the quantities of materials and labor required to complete the project.

Price the quantities: Once the quantities have been measured, the estimator must price the quantities. This involves obtaining quotes from suppliers and subcontractors. The estimator must also consider the cost of equipment and overhead expenses.

Assemble the cost estimate: The estimator must then assemble the cost estimate. This involves grouping the costs by trade and by phase of the project. The estimator must also include a contingency fund to cover the costs of unexpected events.

Challenges of Detailed Cost Estimation
There are a number of challenges associated with detailed cost estimation, including:

Accuracy of the construction drawings: The accuracy of the construction drawings is essential for a detailed cost estimate. If the drawings are inaccurate, the cost estimate will also be inaccurate.

Completeness of the construction drawings: The construction drawings must be complete in order to prepare a detailed cost estimate. If the drawings are incomplete, the estimator may not be able to accurately estimate the cost of all of the work that needs to be done.

Accuracy of material prices and labor rates: Material prices and labor rates can fluctuate over time. This can make it difficult to accurately estimate the cost of the project.

Essential Factors for Improving the Accuracy of Detailed Cost Estimates

Despite the challenges, there are a number of things that can be done to improve the accuracy of detailed cost estimates, including:

Use multiple methods: Using multiple methods to estimate the cost of a project can help to reduce the uncertainty in the estimate.

Use realistic assumptions: When making assumptions about the project scope, design, and construction methods, it is important to be realistic.

Get feedback from experts: It is helpful to get feedback on the cost estimate from other experts, such as experienced cost estimators or contractors.

Update the cost estimate regularly: As the project progresses, it is important to update the cost estimate regularly to reflect changes in the project scope, design, and construction methods.

2.4 Factors Affecting Construction Costs

The cost of a construction project is affected by a wide range of factors. It is important to carefully consider all of these factors when estimating the cost of a project. By doing so, owners and contractors can help to ensure that the project is completed on time and within budget. Some of the factors affecting construction costs include:

Project scope: The scope of the project is the first and foremost factor that affects the cost. It defines the size, complexity, and duration of the project. A larger and more complex project will typically cost more than a smaller and simpler project.

Design: The design of the project also plays a significant role in determining its cost. A more complex design will typically cost more than a simpler design. Additionally, the choice of materials and finishes can also have a significant impact on cost.

Location: The location of the project can also affect its cost. Projects in urban areas tend to cost more than projects in rural areas. This is due to factors such as higher labor costs and land prices.

Market conditions: The prevailing market conditions can also affect the cost of a construction project. For example, if there is a shortage of labor or materials, the cost of the project will likely be higher.

Time of year: The time of year in which the project is constructed can also affect its cost. For example, construction projects in the winter tend to cost more than projects in the summer. This is due to factors such as bad weather conditions and the need to use specialized equipment.

In addition to these general factors, there are a number of other factors that can affect the cost of a construction project, such as:

Site conditions: Difficult site conditions, such as poor soil conditions or the presence of hazardous materials, can increase the cost of a project.

Permits and fees: The cost of permits and fees can also add to the overall cost of a project.

Unexpected events: Unexpected events, such as bad weather or strikes, can also increase the cost of a project.
Conclusion

CHAPTER 3: MEASUREMENT AND TAKEOFF

3.1Techniques for Measuring Building Elements

There are a number of techniques that can be used to measure building elements. The most appropriate technique will depend on the specific element being measured, the required accuracy, and the available resources.

Manual Measurement

Manual measurement is the traditional method of measuring building elements. It involves using measuring tapes, rulers, and other tools to measure the dimensions of the elements. Manual measurement is a relatively simple and inexpensive method, but it can be time-consuming and inaccurate, especially for large or complex elements. To measure a building element using manual measurement, you will need the following tools:

- Measuring tape
- Ruler
- Level
- Pencil
- Paper

Follow these steps to measure a building element:

- Identify the element that you need to measure.
- Determine the dimensions that you need to measure.
- Use the appropriate measuring tool to measure the dimensions.
- Record the measurements on paper.

The following are some factors to consider for manual measurement:

- Make sure that the measuring tools are calibrated and in good condition.

- Take multiple measurements of each dimension and average the results.

- Be careful of obstructions, such as electrical outlets, plumbing fixtures, and furniture.

- Document your work by recording the measurements, the date and time of the measurements, and the person who took the measurements.

Laser Scanning

Laser scanning is a non-contact method of measuring building elements. It uses a laser scanner to create a 3D point cloud of the elements. The point cloud can then be used to measure the dimensions of the elements with high accuracy. Laser scanning is a relatively fast and accurate method, but it can be expensive and requires specialized equipment.

To measure a building element using laser scanning, you will need the following equipment:

- Laser scanner
- Software for processing the scan data

Follow these steps to measure a building element using laser scanning:

- Place the laser scanner in a central location and scan the element.

- Use the software to process the scan data and create a 3D point cloud.

- Measure the dimensions of the element in the 3D point cloud.

The following are some factors to consider for laser scanning:

- Make sure that the laser scanner is calibrated and in good condition.

- Scan the element from multiple locations to ensure complete coverage.

- Use the appropriate software to process the scan data and create a high-quality 3D point cloud.

Photogrammetry

Photogrammetry is a non-contact method of measuring building elements. It uses photographs of the elements to create a 3D model of the elements. The 3D model can then be used to measure the dimensions of the elements. Photogrammetry is a relatively fast and accurate method, and it is becoming less expensive as the technology improves.

To measure a building element using photogrammetry, you will need the following equipment:
- Camera
- Software for processing the photographs

Follow these steps to measure a building element using photogrammetry:

- Take photographs of the element from multiple angles.
- Use the software to process the photographs and create a

3D model of the element.
- Measure the dimensions of the element in the 3D model.

The following are some factors to consider for photogrammetry:

- Take high-resolution photographs of the element.
- Make sure that the photographs are well-lit and in focus.
- Take photographs from multiple angles to ensure complete coverage.
- Use the appropriate software to process the photographs and create a high-quality 3D model.

Drone Surveying

Drone surveying is a non-contact method of measuring building elements. It uses a drone to take photographs of the elements. The photographs can then be used to create a 3D model of the elements. Drone surveying is a relatively fast and inexpensive method, but it is less accurate than laser scanning or photogrammetry.

To measure a building element using drone surveying, you will need the following equipment:

- Drone
- Camera
- Software for processing the photographs

Follow these steps to measure a building element using drone surveying:

- Fly the drone over the element and take photographs.
- Use the software to process the photographs and create a 3D model of the element.
- Measure the dimensions of the element in the 3D model.

Factors to consider for drone surveying:

- Use a drone with a high-resolution camera.
- Make sure that the photographs are well-lit and in focus.

- Take photographs from multiple angles to ensure complete coverage.
- Use the appropriate software to process the photographs and create a high-quality 3D model.

The best way to choose the right technique for measuring a building element is to consider the following factors:

Accuracy: How accurate do the measurements need to be?

Cost: How much money is available for the measurement?

Resources: What resources are available for the measurement, such as equipment and personnel?

Time: How much time is available for the measurement?
Manual measurement is a good option for small and simple elements, where high accuracy is not required. Laser scanning and photogrammetry are good options for large and complex elements, where high accuracy is required. Drone surveying is a good option for large and complex elements, where high accuracy is not required and the element is difficult or dangerous to access.

These are some examples of when to use each technique:

Manual measurement:
- Measuring the dimensions of a door or window
- Measuring the length of a wall
- Measuring the height of a ceiling

Laser scanning:
- Measuring the dimensions of a building
- Measuring the volume of a room
- Measuring the deformation of a bridge

Photogrammetry:
- Measuring the dimensions of a complex object, such as a sculpture or a machine

- Measuring the volume of a stockpile
- Measuring the area of a forest

Drone surveying:
- Measuring the dimensions of a large area, such as a construction site or a farm
- Measuring the volume of a stockpile
- Inspecting a building or structure for defects

3.2 Quantifying Material and Labor Requirements

Quantifying material and labor requirements is the process of estimating the amount of materials and labor required to complete a project. This is an important step in the project planning process, as it allows you to budget for the project and to ensure that you have the necessary resources to complete the project on time and within budget.

There are a number of different methods that can be used to quantify material and labor requirements. The most appropriate method will depend on the project, the available information, and the desired level of accuracy.

Common methods for quantifying material and labor requirements include:

Quantity takeoff: QTO is a method of estimating the amount of materials required to complete a project. It involves measuring the quantities of materials required for each task or activity in the project. QTO is typically performed using the project drawings and specifications.

Assembly pricing: Assembly pricing is a method of estimating the cost of labor and materials required to complete a specific assembly. It involves identifying the individual components of the assembly and estimating the cost of each component. Assembly pricing is typically used for complex assemblies, such as a roof or a wall.

Resource-based estimation: Resource-based estimation is a method of estimating the cost of labor and materials required to complete a project by breaking down the project into smaller tasks or activities and estimating the cost of each task or activity. Resource-based estimation is typically used for projects with a high degree of variability or complexity.

Once the material and labor requirements have been quantified, they can be used to:

Create a budget for the project: The material and labor requirements can be used to create a budget for the project. This will help you to ensure that you have the necessary funds to complete the project.

Develop a project schedule: The material and labor requirements can be used to develop a project schedule. This will help you to identify the critical tasks and to ensure that the project is completed on time.

Track progress during the project: The material and labor requirements can be used to track progress during the project. This will help you to identify any potential problems and to take corrective action as needed.

These are some factors to consider for quantifying material and labor requirements:

Use accurate and up-to-date information: The accuracy of your estimates will depend on the accuracy of the information that you use. Make sure to use the latest project drawings and specifications, and to obtain quotes from suppliers and subcontractors.

Be realistic: When making estimates, it is important to be realistic. Don't underestimate the amount of materials or labor required, or you may end up running out of resources or having to extend the project schedule.

Include a contingency: It is always a good idea to include a contingency in your estimates. This will allow you to cover any unexpected costs or delays.

3.3Leveraging Digital Tools for Efficient Quantity Surveying

Digital tools have revolutionized the quantity surveying profession. By leveraging these tools, quantity surveyors can streamline their workflows, improve the accuracy of their estimates, and collaborate more effectively with other stakeholders.

Some of the most common digital tools used by quantity surveyors include:

Quantity takeoff software: QTO software automates the process of measuring the quantities of materials required to complete a project. This can save quantity surveyors a significant amount of time and effort, and can help to improve the accuracy of their estimates.

Building information modeling (BIM) software: BIM software creates a digital 3D model of a building or structure. This model can be used to perform a variety of tasks, including quantity takeoff, cost estimation, and project scheduling.

Cost estimation software: Cost estimation software helps quantity surveyors to estimate the cost of a project, including the cost of materials, labor, and equipment. Cost estimation software can also be used to track costs during the project and to identify any potential overruns.

Project management software: Project management software helps quantity surveyors to plan, schedule, and track projects. This software can also be used to collaborate with other stakeholders, such as architects, engineers, and contractors.

In addition to these general-purpose tools, there are also a number of specialized digital tools available for quantity surveyors. For example, some software can be used to estimate the cost of specific types of construction projects, such as roads or bridges. Other software can be used to generate reports for specific purposes, such as planning applications or building permits.

These are some specific examples of how digital tools can be used to improve the efficiency of quantity surveying:

Quantity takeoff: QTO software can be used to automate the process of measuring the quantities of materials required to complete a project. This can save quantity surveyors a significant amount of time and effort, and can help to improve the accuracy of their estimates.

Cost estimation: Cost estimation software can help quantity surveyors to estimate the cost of a project, including the cost of materials, labor, and equipment. Cost estimation software can also be used to track costs during the project and to identify any potential overruns.

Collaboration: Digital tools can be used to improve collaboration between quantity surveyors and other stakeholders, such as architects, engineers, and contractors. For example, BIM software can be used to share a digital 3D model of a building or structure with all stakeholders. This can help to reduce errors and omissions and to ensure that everyone is working from the same information.

CHAPTER 4: BILL OF QUANTITIES

4.1 The Structure and Format of a BOQ

A bill of quantities is a document that lists the quantities of materials and labor required to complete a construction project. It is the construction document that provides a detailed breakdown of the materials and labor required to complete a construction project. BOQs are typically prepared by quantity surveyors and used by contractors to estimate the cost of a project and to track costs during construction.

BOQs can be structured in a number of different ways, but they typically include the following information:

- **Project details**: This includes the project name, location, and client details.

- **Scope of work**: This section describes the work that is to be completed under the contract.

- **Bill of quantities**: This section lists the quantities of materials and labor required to complete the work.

- **Pricing**: This section provides the unit prices for the materials and labor listed in the bill of quantities.

- **Total cost**: This section provides the total cost of the project, based on the quantities and unit prices listed in the BOQ.

The format of a BOQ can also vary depending on the project and the specific requirements of the client. However, most BOQs are organized in a tabular format, with the following columns:

- **Item**: This column lists the item number for each item in the BOQ.

- **Description**: This column provides a brief description of each item.

- **Quantity**: This column lists the quantity of each item that is required to complete the project.

- **Unit**: This column specifies the unit of measurement for each item.

- Unit price: This column lists the unit price for each item.

- **Total**: This column calculates the total cost of each item by multiplying the quantity by the unit price.

In addition to the above information, BOQs may also include other information, such as the following:

Specifications: This section provides detailed specifications for the materials and labor to be used on the project.

Terms and conditions: This section outlines the terms and conditions of the contract, such as the payment schedule and the dispute resolution process.

The following are additional information for preparing a BOQ:

- **Comprehensiveness**: The BOQ should include all of the materials and labor required to complete the project, including any indirect costs, such as site overheads and profit.

- **Accuracy**: The quantities and unit prices listed in the BOQ should be as accurate as possible. This will help to ensure that the project is completed on time and within budget.

- **Clarity and conciseness**: The BOQ should be easy to read and understand. Avoid using jargon or technical terms that may not be familiar to all stakeholders.

- **Feedback**: Once the BOQ has been prepared, it is a good idea to get feedback from other stakeholders, such as the architect, engineer, and contractor. This will help to ensure that the BOQ is complete and accurate.

4.2 Preparing a BOQ: Step-by-Step Guide

A Bill of Quantities serves as a crucial component for cost estimation, tendering, and project management. Preparing a BoQ is a detailed process that involves several key steps. The following is a typical guide to help you understand the step-by-step process of creating a Bill of Quantities:

1. Initial Project Evaluation
- Begin by thoroughly understanding the project's scope, including the architectural and engineering drawings, specifications, and any other relevant project documents.
- Determine the project's requirements, objectives, and constraints.
- Identify and clarify any ambiguities or discrepancies in the project documents.

2. Create a Work Breakdown Structure (WBS)
- Develop a systematic WBS that breaks down the project into manageable sections or work packages.
- Organize the WBS in a way that reflects the logical sequence of the construction process, typically starting from site preparation and progressing through various

construction phases.

3. Standardize Measurement Units

- Establish a clear and consistent set of measurement units to be used throughout the BoQ. Standard units typically include square meters (m²), cubic meters (m³), linear meters (lm), and kilograms (kg).
- Ensure that these measurement units align with industry standards and are suitable for the specific types of work in the project.

4. Quantify Each Item

- For each item in the WBS, determine the quantities required based on the project's plans and specifications.
- Ensure that you accurately measure and quantify every aspect of the work, including materials, labor, and equipment.
- Double-check your calculations to minimize errors in quantity estimation.

5. Selection of Materials and Methods

- Specify the materials, components, and construction methods that will be used for each item in the BoQ.
- Ensure that all materials and methods are in compliance with the project's specifications and meet relevant industry standards and regulations.

6. Prepare Descriptions and Specifications

- Provide detailed descriptions and specifications for each item to ensure that contractors have a clear understanding of the work required.
- Include any relevant technical details, quality standards, and other requirements for each item.

7. Apply Rates and Prices

- Assign unit rates and prices to each item based on current market rates and your organization's cost data.
- Rates should include the cost of materials, labor,

equipment, overhead, and profit margins.

8. Organize and Format
- Organize the BoQ in a clear and standardized format, typically following a standard template or format used in the construction industry.
- Use a logical sequence that matches the construction process, ensuring that items are presented in a structured manner.

9. Review and Quality Control
- Conduct a thorough review of the entire BoQ to verify the accuracy of measurements, descriptions, rates, and prices.
- Ensure consistency and completeness throughout the document.
- Address any inconsistencies, errors, or discrepancies.

10. Document Revisions and Updates:
- Keep the BoQ up to date throughout the project. If there are changes in project scope, materials, or methods, make appropriate revisions to the BoQ and communicate these changes to all relevant parties.

11. Tendering and Procurement
- Share the BoQ with contractors, suppliers, or subcontractors during the tendering process.
- Use the BoQ as a basis for soliciting bids and proposals from potential construction and trade professionals.

12. Construction and Cost Control
- Use the BoQ as a reference for cost control throughout the construction project.
- Compare actual costs with the estimated costs from the BoQ and address any discrepancies or variations.

13. Project Completion

- At the end of the project, the BoQ is used for reconciling final accounts, making payments, and evaluating project performance.

4.3 Real-Life Examples and Sample BOQ Template

Real-Life Examples:

- A BOQ for a new house might include items such as concrete, bricks, timber, roofing materials, windows, doors, electrical wiring, plumbing pipes, and sanitary fittings.

- A BOQ for a commercial building might include items such as steel beams, concrete slabs, glass curtain walls, elevators, and air conditioning units.

- A BOQ for a road construction project might include items such as asphalt, gravel, concrete, steel reinforcement bars, and drainage pipes. Table 1 presents a sample BOQ template for a new house.

Table 1: Sample BOQ template for a new house

Item	Description	Quantity	Unit	Unit price	Total
1	Concrete	100 cubic meters	m3	$100	$10,000
2	Bricks	10,000	bricks	$0.50	$5,000
3	Timber	20 cubic meters	m3	$150	$3,000
4	Roofing materials	100 square meters	m2	$20	$2,000
5	Windows	10	windows	$100	$1,000
6	Doors	5	doors	$200	$1,000
7	Electrical	100	m	$10	$1,000

	wiring	meters			
8	Plumbing pipes	50 meters	m	$5	$250
9	Sanitary fittings	10	fittings	$10	$100
10	Total				$22,650

CHAPTER 5: COST CONTROL AND BUDGETING

5.1 The Crucial Role of Cost Control

Cost control is the process of identifying, measuring, and managing the costs associated with a project. It is an essential part of any project management process, as it helps to ensure that the project is completed on time and within budget.

Cost control is crucial in construction projects because of the following reasons:

The high cost of construction projects: Construction projects are typically very expensive, so it is important to control costs in order to avoid financial losses.

The complex nature of construction projects: Construction projects are complex and involve a variety of different stakeholders, such as architects, engineers, contractors, and suppliers. This complexity can make it difficult to track costs and identify potential cost overruns.

The dynamic nature of construction projects: Construction projects are dynamic and can be affected by a variety of factors, such as weather conditions, unforeseen site conditions, and changes to the project scope. This can make it difficult to budget accurately and control costs. Effective cost control can help construction projects to achieve the following benefits:

Improved profitability: By controlling costs, construction companies can increase their profits.

Enhanced competitiveness: Construction companies that can control their costs are more competitive in the marketplace.

Reduced risk: Cost control helps to identify and manage risks, such as the risk of cost overruns and the risk of project delays.

Improved efficiency and productivity: Cost control helps to identify and eliminate waste, which can improve efficiency and productivity.

Enhanced customer satisfaction: By completing projects on time and within budget, construction companies can improve customer satisfaction.

There are a number of different cost control techniques that can be used in construction projects. Some of the most common techniques include:

Budgeting: Budgeting is the process of creating a plan for how much money will be spent on a project. Budgets should be realistic and achievable, and they should be updated regularly as the project progresses.

Tracking: Tracking is the process of monitoring actual costs against the budget. This helps to identify any areas where costs are overrunning and allows corrective action to be taken early on.

Forecasting: Forecasting is the process of estimating future costs based on past performance. This helps to identify any potential cost overruns and to develop contingency plans.

Value engineering: Value engineering is a process that helps to identify and eliminate unnecessary costs without sacrificing quality or functionality.

Risk management: Risk management is the process of identifying, assessing, and mitigating risks. Cost overruns are a common risk in construction projects, so it is important to have a risk management plan in place.

Effective cost control is essential for the success of any construction project. By following the tips above, construction companies can improve their profitability, enhance their competitiveness, reduce risk, improve efficiency and productivity, and enhance customer satisfaction.

The factors below are some additional measures to consider for effective cost control in construction projects:

Use project management software: Project management software can help construction companies track costs, budget, and forecast future costs.

Communicate effectively with stakeholders: It is important to communicate effectively with all stakeholders about the cost control process. This will help to ensure that everyone is aware of their roles and responsibilities and that everyone is working towards the same goals.

Be proactive: It is important to be proactive and to identify potential cost overruns early on. This will allow you to take corrective action and minimize the impact on the project.

5.2 Budgeting for Construction Projects

Budgeting is the process of creating a plan for how much money will be spent on a project. It is an essential part of any project management process, and it is especially important in construction projects, where the costs can be high and complex. To create a detailed budget for a construction project, you can follow these steps:

- Identify the scope of work. What work needs to be completed? What materials and labor will be required? Be as specific as possible.

- Develop a schedule for the project. When will the project start and finish? What are the key milestones?

- Identify the risks associated with the project. What are the potential risks that could impact the project budget? For example, you may need to consider the risk of bad weather, unforeseen site conditions, or changes to the project scope.
- Estimate the cost of each task. This includes the cost of materials, labor, equipment, and any other associated costs.

- Be sure to get quotes from multiple suppliers and contractors to get the best possible prices.

- Allocate the costs to different budget categories. Some common budget categories include:

❖ Site preparation
❖ Foundation
❖ Framing
❖ Exterior finishes
❖ Interior finishes
❖ MEP systems

Calculate the total project budget. Be sure to include a contingency fund in the budget to cover any unexpected costs.

These are some considerations for developing a detailed budget:

- Use a project management software program. This can help you to track costs, budget, and forecast future costs.

- Break the project down into smaller tasks. This will make it easier to estimate the cost of each task and to identify potential cost overruns.

- Get quotes from multiple suppliers and contractors. This will help you to get the best possible prices and to minimize the risk of overpaying.

- Be realistic when estimating costs. Don't underestimate the cost of materials and labor, or you may end up running out of money before the project is finished.

- Be flexible. Things don't always go according to plan in construction projects, so it is important to be flexible with your budget. Be prepared to adjust your budget as needed to account for unexpected costs or changes to the project scope.

- Track progress regularly. Once you have created a budget, it is important to track your progress regularly. This will help you to identify any areas where costs are overrunning and allow you to take corrective action early on.

- Communicate with stakeholders. It is important to communicate effectively with all stakeholders about the budget. This will help to ensure that everyone is aware of the budget and that everyone is working towards the same goals.

With these steps, you can create a detailed budget that will help you to complete your construction project on time and within budget.

The following are some additional factors to consider for developing a detailed budget for specific construction projects:

Residential construction: For residential construction projects, you may want to consider budgeting for the following items:

- Permits and fees
- Site preparation
- Foundation
- Framing
- Exterior finishes

- Interior finishes
- MEP systems
- Landscaping

Commercial construction: For commercial construction projects, you may want to consider budgeting for the following items:
- Permits and fees
- Site preparation
- Foundation
- Framing
- Exterior finishes
- Interior finishes
- MEP systems
- HVAC
- Fire protection
- Security systems

Infrastructure construction: For infrastructure construction projects, you may want to consider budgeting for the following items:

- Permits and fees
- Site preparation
- Earthwork
- Drainage
- Paving
- Utilities
- Bridges and tunnels

Once you have developed a detailed budget, it is important to review it regularly and make adjustments as needed. This will help you to stay on track and to avoid any budget surprises.

5.3 Effective Cost Monitoring and Reporting

Effective cost monitoring and reporting are essential for any construction project. By tracking costs and identifying areas where costs are overrunning, project managers can take corrective action early on and avoid financial losses.

There are a number of different ways to monitor costs on a construction project. One common method is to use a cost management software program. These programs can help project managers to track costs, budget, and forecast future costs.

Another common method for cost monitoring is to use a cost control plan. A cost control plan is a document that outlines the project budget and identifies the key milestones and deliverables. The cost control plan also includes a process for tracking costs and identifying potential cost overruns.

Project managers should also use regular reporting to track costs and identify areas where costs are overrunning. Some common cost reports include:

Budget vs. actual report: This report compares the project budget to the actual costs incurred to date.

Earned value management report: This report tracks the progress of the project and compares the value of the work completed to the budget and schedule.

Variance report: This report identifies any areas where costs are overrunning or underrunning the budget.

By monitoring costs and using regular reporting, project managers can identify potential problems early on and take corrective action to avoid financial losses. The following factors can benefit effective cost monitoring and reporting:

- Use a cost management software program. This can help you to track costs, budget, and forecast future costs.

- Develop a cost control plan. This document should outline the project budget, identify the key milestones and deliverables, and include a process for tracking costs and identifying potential cost overruns.

- Use regular reporting to track costs and identify areas where costs are overrunning. Some common cost reports include the budget vs. actual report, the earned value management (EVM) report, and the variance report.

- Communicate with stakeholders. It is important to communicate effectively with all stakeholders about the cost monitoring and reporting process. This will help to ensure that everyone is aware of the project budget and that everyone is working towards the same goals.

Project managers can use the above guide to effectively monitor costs and report on project progress to stakeholders.

CHAPTER 6:
TENDERING AND
PROCUREMENT

6.1Understanding
Procurement Methods

Procurement methods are the processes and procedures that organizations use to acquire goods and services from external suppliers. There are a variety of different procurement methods available, each with its own advantages and disadvantages. The most common procurement methods include:

Open tendering: Open tendering is a competitive bidding process that is open to all qualified suppliers. This method is typically used for large or complex projects, where it is important to get the best possible price and quality.

Restricted tendering: Restricted tendering is similar to open tendering, but only a select group of pre-approved suppliers are invited to bid. This method is typically used for projects where the organization has a specific supplier in mind, but wants to get multiple quotes to ensure that they are getting the best possible price.

Request for proposal (RFP): An RFP is a document that is sent to potential suppliers requesting them to submit proposals for a specific project. RFPs are typically used for complex projects where the organization needs a lot of information from the supplier in order to make a decision.

Request for quotation (RFQ): An RFQ is a document that is sent to potential suppliers requesting them to submit quotes for a specific product or service. RFQs are typically used for simple projects or for the purchase of standard items.

Single sourcing: Single sourcing is the process of acquiring goods or services from a single supplier. This method is typically used for projects where the organization has a long-term relationship with the supplier or where the supplier has a unique product or service that is not available from other suppliers.

The best procurement method for a particular project will depend on a number of factors, including the size and complexity of the project, the budget, and the organization's procurement policies and procedures. Table 2 summarizes the key advantages and disadvantages of each procurement method.

Table 2: Advantages and disadvantages of the procure

ment methods

Procurement method	Advantages	Disadvantages
Open tendering	Competitive process, transparent, open to all qualified suppliers	Can be time-consuming and expensive, difficult to compare bids
Restricted tendering	More efficient than open tendering, allows organization to select from a pre-approved list of suppliers	May not be as competitive as open tendering, can lead to favoritism
Request for proposal (RFP)	Allows organization to get detailed information from suppliers, good for complex projects	Can be time-consuming and expensive to prepare and evaluate RFPs
Request for quotation (RFQ)	Simple and efficient process, good for the purchase of standard items	May not be as competitive as other procurement methods, difficult to compare quotes

Single sourcing	Can be quick and efficient, good for projects where there is a long-term relationship with the supplier or where the supplier has a unique product or service	Can lead to higher prices, less competition, and less innovation

Organizations should carefully consider the advantages and disadvantages of each procurement method before selecting a method for a particular project.

6.2 Preparing Comprehensive Tender Documentation

Comprehensive tender documentation is essential for ensuring a fair and competitive tendering process. The tender documentation should provide all of the information that suppliers need to prepare and submit accurate and complete bids. The following are some considerations for preparing comprehensive tender documentation:

Be clear and concise: The tender documentation should be clear and concise, and it should be easy for suppliers to understand. Avoid using jargon or technical terms that may not be familiar to all suppliers.

Be comprehensive: The tender documentation should include all of the information that suppliers need to prepare and submit accurate and complete bids. This includes the scope of work, the requirements for the goods or services, the evaluation criteria, and the submission instructions.

Be fair and impartial: The tender documentation should be fair and impartial, and it should not give any unfair advantage to any one supplier.

Be realistic: The requirements for the goods or services should be realistic and achievable.

Be flexible: The tender documentation should be flexible enough to allow suppliers to submit innovative and cost-effective solutions.

The following is a checklist of items that should typically be included in tender documentation:

Invitation to tender: This document formally invites suppliers to bid on the project. It should include information about the project, the tendering process, and the submission deadline.

Instructions to tenderers: This document provides detailed instructions on how to prepare and submit a bid. It should include information about the evaluation criteria, the format of the bid, and the submission requirements.

Scope of work: This document describes the work that is to be completed under the contract. It should be as detailed as possible, and it should include any specific requirements that the organization has.

Technical specifications: This document provides detailed technical specifications for the goods or services that are being procured. It should be as detailed as possible, and it should include any specific requirements that the organization has.

Form of contract: This document is the contract that will be signed between the organization and the successful supplier. It should be drafted by a qualified lawyer, and it should include all of the terms and conditions of the contract.

In addition to the above items, the tender documentation may also include other items, such as:

Bill of quantities: This document lists the quantities of goods or services that are being procured.

Pricing schedule: This document provides the supplier with a template for pricing their bid.

Evaluation criteria: This document describes the criteria that will be used to evaluate the bids.

Submission requirements: This document specifies how the bids should be submitted, such as by email, by post, or in person.

Organizations can consider the above to prepare comprehensive tender documentation that will help them to conduct a fair and competitive tendering process.

6.3 Evaluating Bids and Awarding Contracts

Once bids have been received, organizations need to evaluate the bids and award the contract to the successful supplier. The bid evaluation process should be fair and impartial, and it should be based on the evaluation criteria that were outlined in the tender documentation. The following are some steps involved in evaluating bids:

Check for completeness: The first step is to check that all of the required information has been submitted by each supplier. This includes checking that the bid is in the correct format, that all of the required documentation has been submitted, and that the bid is complete.

Evaluate the bids: The next step is to evaluate the bids based on the evaluation criteria that were outlined in the tender documentation. This may involve scoring the bids against each criterion or ranking the bids from best to worst.

Identify the winning bid: The winning bid is the bid that has the highest score or ranking, based on the evaluation criteria.

Award the contract: The final step is to award the contract to the successful supplier. This typically involves sending a letter of award to the supplier and signing a contract.

The following can help in evaluating bids and awarding contracts:

Use a clear and objective evaluation process: The bid evaluation process should be clear and objective, and it should be based on the evaluation criteria that were outlined in the tender documentation.

Be fair and impartial: The bid evaluation process should be fair and impartial, and it should not give any unfair advantage to any one supplier.

Document the evaluation process: The bid evaluation process should be documented, and the documentation should be kept on file for future reference.

Award the contract to the winning bidder: The contract should be awarded to the supplier that submitted the winning bid, based on the evaluation criteria.

CHAPTER 7: VARIATIONS AND CHANGE ORDERS

7.1 Managing Changes in Construction Projects

Change is inevitable in construction projects. Unforeseen site conditions, changes in the project scope, and unexpected delays can all lead to changes in the project schedule and budget. Effective change management is essential for ensuring that construction projects are completed on time and within budget and that the project meets the needs of the client.

Establish a change management process
The first step is to establish a change management process. This process should outline how changes will be identified, evaluated, approved, and implemented. The change management process should be documented and communicated to all stakeholders.

Identify and evaluate changes
Changes should be identified and evaluated as early as possible. This will help to minimize the impact of the changes on the project schedule and budget. Changes should be evaluated based on their impact on the project scope, schedule, budget, and quality.

Approve changes

Changes should be approved by the appropriate stakeholders. This may include the client, the project manager, the architect, the engineer, and the contractor.

Implement changes
Once a change has been approved, it should be implemented as quickly as possible. This will help to minimize the impact of the change on the project schedule and budget.

Additional factors to consider

Communicate effectively: It is important to communicate effectively with all stakeholders about changes to the project. This will help to ensure that everyone is aware of the changes and that everyone is working towards the same goals.

Be proactive: It is important to be proactive and to identify potential changes early on. This will allow you to take steps to minimize the impact of the changes.

Be flexible: Things don't always go according to plan in construction projects, so it is important to be flexible and to adapt to changes as needed.
Examples of common changes in construction projects

Changes to the project scope: This may include changes to the design of the project, the materials to be used, or the work to be completed.

Changes to the project schedule: This may be caused by unforeseen site conditions, delays in obtaining permits, or bad weather.
Changes to the project budget: This may be caused by changes to the project scope, changes to the project schedule, or unexpected costs.

7.2 Valuing Variations Accurately
Valuing variations accurately is essential for ensuring that

construction projects are completed on time and within budget. Variations are changes to the scope of work of a construction project. They can be caused by a variety of factors, such as unforeseen site conditions, changes in the project scope, and unexpected delays.

There are a number of different methods for valuing variations. The most common method is to use the contract rate. The contract rate is the rate that the contractor has agreed to do the work for in the contract. However, the contract rate may not be appropriate for all variations. For example, if the variation is for work that is not covered by the contract, then the contract rate will not be applicable.

In some cases, it may be necessary to use a different method for valuing variations. For example, if the variation is for work that is more complex than the work that is covered by the contract, then it may be necessary to use a daywork rate. A daywork rate is a rate that is charged per day for labor and equipment.

It is important to note that there is no one-size-fits-all approach to valuing variations. The best method for valuing a variation will depend on the specific circumstances of the variation. The following are some considerations for valuing variations accurately:

Identify the scope of the variation: The first step is to identify the scope of the variation. This includes identifying the work that needs to be completed, the materials that need to be used, and the labor that will be required.

Estimate the cost of the variation: Once you have identified the scope of the variation, you can estimate the cost of the variation. This includes estimating the cost of materials, labor, and equipment.

Compare the estimated cost to the contract rate: If the variation is for work that is covered by the contract, then you can compare the estimated cost to the contract rate. If the estimated cost is higher than the contract rate, then you will need to negotiate a new price with the contractor.

Use a different method for valuing variations if necessary: If the variation is for work that is not covered by the contract, or if the variation is for work that is more complex than the work that is covered by the contract, then you may need to use a different method for valuing the variation. For example, you may need to use a daywork rate.

It is important to get the agreement of the client and the contractor before implementing any variations. This will help to avoid disputes later on. You can also consider the following for valuing variations accurately:

Be realistic: When estimating the cost of a variation, be realistic about the cost of materials, labor, and equipment.

Be transparent: Be transparent with the client and the contractor about the cost of the variation.

Be flexible: Things don't always go according to plan in construction projects, so be prepared to adjust the value of a variation as needed.

7.3 Negotiating and Administering Change Orders

A change order is a written document that authorizes a change to the scope of work, schedule, or budget of a construction project. Change orders are common in construction projects, and they can be caused by a variety of factors, such as unforeseen site conditions, changes in the project scope, and unexpected delays.

Negotiating and administering change orders effectively is essential for ensuring that construction projects are completed on time and within budget.

Negotiating change orders

When negotiating a change order, it is important to consider the following factors:

The scope of the change: What work needs to be completed? What materials and labor will be required?

The impact of the change on the project schedule and budget: How will the change impact the project schedule and budget? Will the change require additional time or money?

The fairness of the price: Is the price that the contractor is proposing for the change fair and reasonable?

It is also important to get the agreement of the client and the contractor before implementing any change orders. This will help to avoid disputes later on. The following are some factors to consider for negotiating change orders effectively:

Be prepared: Before negotiating a change order, be prepared with information about the scope of the change, the impact of the change on the project schedule and budget, and a fair and reasonable price for the change.

Be reasonable: Be reasonable in your negotiations and be willing to compromise.

Be clear and concise: Be clear and concise in your communication with the client and the contractor.

Document everything: Document everything in writing, including the scope of the change, the impact of the change on the project schedule and budget, and the agreed-upon price for the change.

Administering change orders

Once a change order has been negotiated and agreed upon, it is important to administer it effectively. This includes:

Updating the project schedule and budget: The project schedule and budget should be updated to reflect the change order.

Issuing a written change order: A written change order should be issued to the contractor. The change order should include the scope of the change, the impact of the change on the project schedule and budget, and the agreed-upon price for the change.

Monitoring the progress of the change order: The progress of the change order should be monitored to ensure that it is completed on time and within budget.

Making payments to the contractor: The contractor should be paid for the work completed under the change order in accordance with the terms of the contract.

Change orders can also be administered considering the following:

Be organized: Keep track of all change orders and their associated documentation.

Be communicative: Communicate regularly with the client and the contractor about the change orders.

Be flexible: Things don't always go according to plan in construction projects, so be prepared to adjust the administration of change orders as needed.

CHAPTER 8: FINAL ACCOUNTS AND PAYMENT

8.1Interim and Final Payments

Interim and final payments are essential parts of construction projects. Interim payments are regular payments that are made to the contractor during the course of the project, while the final payment is made once the project is complete and all of the work has been accepted by the client.

The process for making interim and final payments is typically outlined in the construction contract. However, there are some general principles that apply to all interim and final payments.

Interim payments

Interim payments are typically made based on the percentage of work that has been completed. For example, the contract may specify that the contractor will receive 10% of the total contract price upon completion of the foundation, 20% upon completion of the framing, and so on.

The contractor is typically required to submit a payment application to the client before receiving an interim payment. The payment application should include a detailed breakdown of the work that has been completed and the cost of the work.

The client will typically review the payment application and make adjustments as needed. Once the payment application has been approved, the client will issue a payment to the contractor.

Final payment

The final payment is made once the project is complete and all of the work has been accepted by the client. The contractor is typically required to submit a final payment application to the client before receiving the final payment.

The final payment application should include a detailed breakdown of all of the work that has been completed and the cost of the work. The client will typically review the final payment application and make adjustments as needed. Once the final payment application has been approved, the client will issue a final payment to the contractor. This is a breakdown of the interim and final payment process:

- The contractor submits a payment application to the client. The payment application should include a detailed breakdown of the work that has been completed and the cost of the work.

- The client reviews the payment application and makes adjustments as needed. The client may also require the contractor to provide additional documentation, such as invoices from suppliers or timesheets from employees.

- Once the payment application has been approved, the client issues a payment to the contractor. The payment may be made by check, wire transfer, or other electronic means.

- The contractor repeats steps 1-3 until the project is complete.

- Once the project is complete, the contractor submits a final payment application to the client. The final payment application should include a detailed breakdown of all of the work that has been completed and the cost of the work.

- The client reviews the final payment application and makes adjustments as needed. The client may also require the contractor to provide additional documentation, such as a certificate of completion from the architect or engineer.

- Once the final payment application has been approved, the client issues a final payment to the contractor. The final payment typically includes any remaining balance due to the contractor, as well as any retention money that has been held back.

8.2 Certificates and Valuation in Quantity Surveying

Certificates and valuation are essential parts of quantity surveying. Certificates are documents that confirm that the contractor has completed certain stages of the work or that certain conditions have been met. Valuations are estimates of the value of the work that has been completed.

Certificates

There are a number of different types of certificates that are used in quantity surveying. Some of the most common types of certificates include:

Interim payment certificate: This certificate is issued to the contractor to authorize payment for the work that has been completed to date.

Completion certificate: This certificate is issued to the contractor to certify that the project has been completed in accordance with the contract documents.

Retention release certificate: This certificate is issued to the contractor to release the retention money that has been held back.

Valuation

Valuation is the process of estimating the value of the work that has been completed. Valuations are typically used to determine the amount of interim payments that are due to the contractor.

There are a number of different methods that can be used to value work. The most common method is to use the contract rate. The contract rate is the rate that the contractor has agreed to do the work for in the contract.

However, the contract rate may not be appropriate for all valuations. For example, if the contractor has incurred additional costs due to unforeseen site conditions, then the contract rate may not be fair.

In some cases, it may be necessary to use a different method for valuation. For example, if the variation is for work that is not covered by the contract, then it may be necessary to use a daywork rate. A daywork rate is a rate that is charged per day for labor and equipment. The following are explanation of the certificates and valuation process:

- The contractor submits a payment application to the quantity surveyor. The payment application should include a detailed breakdown of the work that has been completed and the cost of the work.

- The quantity surveyor reviews the payment application and prepares a valuation. The valuation is an estimate of the value of the work that has been completed.

- The quantity surveyor compares the valuation to the contract rate and makes adjustments as needed. For example, if the contractor has incurred additional costs due to unforeseen site conditions, the quantity surveyor may adjust the valuation upwards.

- The quantity surveyor issues an interim payment

certificate to the contractor. The interim payment certificate authorizes payment for the work that has been completed to date.

- The contractor repeats steps 1-4 until the project is complete.

- Once the project is complete, the quantity surveyor prepares a final valuation. The final valuation is an estimate of the value of all of the work that has been completed.

- The quantity surveyor compares the final valuation to the contract rate and makes adjustments as needed. For example, if the contractor has incurred additional costs due to unforeseen site conditions, the quantity surveyor may adjust the final valuation upwards.

- The quantity surveyor issues a completion certificate to the contractor. The completion certificate certifies that the project has been completed in accordance with the contract documents.

- The quantity surveyor issues a retention release certificate to the contractor. The retention release certificate releases the retention money that has been held back.

The following factors should be considered for managing certificates and valuation:

Be clear and concise: The payment application, valuation, and certificates should be clear and concise, and they should be easy for the client to understand.

Be accurate: The payment application, valuation, and certificates should be accurate and complete.

Be timely: The payment application, valuation, and certificates should be issued to the client on time.

Be communicative: Communicate regularly with the client about certificates and valuation. This will help to avoid any disputes or delays.

8.3Successfully Closing
Out Final Accounts

Successfully closing out final accounts is an important step in any construction project. It is the process of reconciling all of the financial transactions associated with the project and ensuring that all of the parties involved are satisfied with the outcome. To successfully close out final accounts, it is important to:

Start early: It is important to start planning for the final account closure process early in the project. This will give you enough time to gather all of the necessary documentation and to resolve any disputes that may arise.

Be organized: Keep track of all of the financial transactions associated with the project, including invoices, receipts, and change orders. This will make it easier to reconcile the final accounts.

Communicate regularly: Communicate regularly with the client and the contractor about the final account closure process. This will help to ensure that everyone is on the same page and that there are no surprises.

Be flexible: Things don't always go according to plan in construction projects, so be prepared to be flexible during the final account closure process. For example, you may need to negotiate with the client or the contractor if there are any disagreements about the final account.

This is a more detailed breakdown of the final account closure process:

- Gather all of the necessary documentation. This

includes invoices, receipts, change orders, and any other documentation that supports the final account.

- Reconcile the final accounts. This involves comparing the contractor's invoice to the client's budget and to the contract documents.

- Review the final accounts with the client and the contractor. This is an opportunity to resolve any disputes and to ensure that everyone is satisfied with the final account.

- Issue the final payment. This is the final payment that is made to the contractor for the work that has been completed.

The following are some additional considerations for successfully closing out final accounts:

- Use standard forms and templates. There are a number of standard forms and templates available that can be used to help with the final account closure process.

- Get professional help. If you are unsure about any aspect of the final account closure process, consider seeking professional help from a quantity surveyor or accountant.

- Be patient. Closing out final accounts can be a time-consuming process, so be patient and persistent.

CHAPTER 9: LEGAL AND ETHICAL CONSIDERATIONS

9.1 The Legal Framework of Quantity Surveying

Quantity surveying is a profession that is governed by a set of laws and regulations, known as the legal framework. This framework includes both domestic and international laws, as well as the codes of ethics and professional conduct that are promulgated by professional bodies. The following are some of the key elements of the legal framework of quantity surveying:

Contracts: Contracts are the foundation of the legal relationship between a quantity surveyor and their client. It is important to ensure that contracts are clear, comprehensive, and adequately protect the interests of both parties.

Professional negligence: Quantity surveyors have a duty of care to their clients. This means that they must act with reasonable skill and care when performing their duties. If a quantity surveyor breaches their duty of care, they may be liable for professional negligence.

Dispute resolution: If a dispute arises between a quantity surveyor and their client, there are a number of dispute resolution options available, including mediation, arbitration, and litigation.

In addition to the above, quantity surveyors may also be subject to other laws and regulations, depending on the jurisdiction in which they are practicing. For example, quantity surveyors are subjected to the Act that regulate the profession, depending on the country. It is important for quantity surveyors to be familiar with the legal framework that applies to their practice. This will help them to avoid disputes and to protect their interests. Quantity surveyors should also consider the following regarding the legal framework of their profession:

Keep records up to date: Quantity surveyors are required to keep accurate records of their work. This includes keeping copies of all contracts, invoices, and other documentation.

Have contracts reviewed by a lawyer: Before signing any contract, it is a good idea to have it reviewed by a lawyer. This will help to ensure that the contract is fair and that it adequately protects your interests.

Get professional liability insurance: Professional liability insurance can protect you from financial losses in the event that you are sued for professional negligence.

Stay up to date on the latest laws and regulations: The laws and regulations that govern the quantity surveying profession can change over time. It is important to stay up to date on the latest changes so that you can ensure that you are in compliance.

9.2 Navigating Professional Ethics and Standards

Quantity surveyors have a responsibility to uphold the highest ethical and professional standards. This is essential for maintaining public trust and ensuring that construction projects are completed fairly and honestly.

The following are some of the key ethical and professional standards that quantity surveyors should adhere to:

Honesty and integrity: Quantity surveyors must be honest and truthful in all of their dealings. They must avoid any conflicts of interest and must act in the best interests of their clients.

Competence and diligence: Quantity surveyors must have the necessary skills and knowledge to carry out their duties competently. They must also be diligent in their work and must take all reasonable steps to avoid errors and omissions.

Objectivity and independence: Quantity surveyors must be objective and independent in their judgment. They should not allow their personal interests or the interests of any third party to influence their decisions.

Professionalism and courtesy: Quantity surveyors should treat their clients, colleagues, and other stakeholders with respect and courtesy. They should also maintain a high level of professionalism in their conduct.

Quantity surveyors may face a number of ethical challenges in the course of their work. For example, they may be asked to provide advice that is beneficial to their client but could potentially harm the public interest. Or, they may be pressured to sign off on work that is not complete or that does not meet the required standards.

In such situations, it is important for quantity surveyors to remember their ethical and professional obligations. They should always act in the best interests of their clients and the public, and they should never compromise their integrity. These are some measures to consider for quantity surveyors on how to navigate professional ethics and standards:

- Be familiar with the relevant codes of ethics and professional conduct. The professional bodies that represent quantity surveyors typically have codes of ethics and professional conduct in place. It is important for quantity surveyors to be familiar with these codes and to adhere to them.

- Seek advice from colleagues and mentors if you are unsure about an ethical issue. If you are unsure about how to handle an ethical issue, it is a good idea to seek advice from colleagues or mentors who have more experience.

- Be prepared to stand up for your principles. There may be times when you are pressured to do something that you believe is unethical. In such situations, it is important to be prepared to stand up for your principles and to refuse to compromise your integrity.

9.3 Dispute Resolution in Quantity Surveying

Dispute resolution in quantity surveying is the process of resolving disagreements between quantity surveyors and their clients, or between quantity surveyors and other stakeholders in a construction project. Disputes can arise over a variety of issues, such as fees, contract interpretation, and the quality of work. There are a number of different dispute resolution mechanisms that can be used in quantity surveying. The most common methods include:

Negotiation: Negotiation is the most common form of dispute resolution in quantity surveying. It involves the parties to the dispute trying to reach an agreement directly with each other.

Mediation: Mediation is a process in which a neutral third party helps the parties to the dispute to reach an agreement. The mediator does not make any decisions on behalf of the parties, but rather helps them to communicate and to explore options for resolving the dispute.

Arbitration: Arbitration is a process in which a neutral third party (the arbitrator) makes a binding decision on the dispute. Arbitration is often used in quantity surveying disputes because it is a relatively quick and inexpensive process.

Litigation: Litigation is the process of resolving a dispute through the courts. Litigation is the most formal and expensive form of dispute resolution, and it should be avoided whenever possible. The best dispute resolution mechanism for a particular case will depend on the specific circumstances of the dispute. It is important to seek legal advice to determine the best course of action.

Quantity surveyors can avoid disputes considering the following:

- Have clear and comprehensive contracts in place. Contracts should clearly outline the scope of work, the fees, and the dispute resolution process.

- Communicate regularly with your clients and other stakeholders. Keep your clients and other stakeholders informed of your progress and any potential problems.

- Be reasonable and flexible. Be willing to negotiate and to compromise in order to resolve disputes quickly and amicably.

CHAPTER 10: FUTURE TRENDS IN QUANTITY SURVEYING

10.1Embracing Digital Transformation in Quantity Surveying

Digital transformation in quantity surveying is the process of adopting new digital technologies and processes to improve the efficiency and effectiveness of quantity surveying services. There are a number of different digital technologies that can be used in quantity surveying, such as:

Building information modeling: BIM is a process of creating and managing digital representations of physical and functional characteristics of places. BIM can be used to improve the accuracy and efficiency of quantity surveying tasks such as taking off quantities and preparing cost estimates.

Artificial intelligence (AI): AI can be used to automate a variety of quantity surveying tasks, such as data entry, document review, and risk assessment.

Machine learning (ML): ML can be used to develop predictive models that can be used to improve the accuracy of quantity surveying estimates.

Blockchain: Blockchain can be used to create secure and

transparent records of quantity surveying data.

Drones: Drones can be used to collect aerial imagery of construction sites, which can be used to take off quantities and monitor progress.

Digital transformation can offer a number of benefits to quantity surveying firms, including:

Improved efficiency and accuracy: Digital technologies can automate many of the repetitive and time-consuming tasks associated with quantity surveying, which can free up quantity surveyors to focus on more strategic tasks. Digital technologies can also help to improve the accuracy of quantity surveying estimates.

Reduced costs: Digital technologies can help to reduce the costs of quantity surveying services by automating tasks and improving efficiency.

Enhanced collaboration: Digital technologies can facilitate collaboration between quantity surveyors and other stakeholders in construction projects, such as architects, engineers, and contractors.

Improved decision-making: Digital technologies can provide quantity surveyors with access to real-time data and insights, which can help them to make better decisions.

Quantity surveying firms that embrace digital transformation can gain a competitive advantage and better meet the needs of their clients. These are some measures for quantity surveying firms on how to embrace digital transformation:

- Start by assessing your current digital maturity. This will help you to identify the areas where you need to make improvements.

- Develop a digital transformation strategy. Your strategy should identify the digital technologies that you want to adopt and the steps that you need to take to implement them.

- Invest in training for your staff. It is important to ensure that your staff have the skills and knowledge to use new digital technologies.

- Start small and scale up over time. Don't try to implement too many new digital technologies at once. Start by implementing a few key technologies and then scale up over time as you gain experience.

10.2 Sustainability and Green Construction Practices

Sustainability and green construction practices are becoming increasingly important in the construction industry. Quantity surveyors can play a vital role in promoting sustainability and green construction practices by:

- Advising clients on sustainable design and construction options. Quantity surveyors can help clients to understand the benefits and costs of different sustainable design and construction options. This can help clients to make informed decisions that align with their sustainability goals.
- Preparing sustainable cost estimates. Quantity surveyors can prepare cost estimates that take into account the environmental and social costs of construction projects. This can help clients to make more informed decisions about their projects.

- Managing sustainable procurement. Quantity surveyors can help clients to procure sustainable materials and products. This can help to reduce the environmental impact of construction projects.

- Monitoring and reporting on sustainable performance.

Quantity surveyors can help clients to monitor and report on the sustainable performance of their construction projects. This can help clients to identify areas where they can improve their sustainability performance.

These are some specific examples of how quantity surveyors can promote sustainability and green construction practices:

- Advise clients on the use of sustainable materials and products. Quantity surveyors can help clients to identify and select sustainable materials and products, such as recycled materials and energy-efficient products.

- Recommend green building certification programs. Quantity surveyors can help clients to select and participate in green building certification programs, such as LEED and BREEAM. These programs assess the environmental performance of buildings and provide certification to buildings that meet certain standards.

- Encourage the use of renewable energy sources. Quantity surveyors can help clients to identify and implement renewable energy sources for their construction projects, such as solar and wind power.

- Promote sustainable site planning. Quantity surveyors can help clients to develop sustainable site plans that minimize the environmental impact of construction projects.

- Encourage water conservation measures. Quantity surveyors can help clients identify and implement water conservation measures for their construction projects, such as rainwater harvesting and greywater reuse.

10.3 Emerging Roles and Opportunities for Quantity Surveyors

The quantity surveying profession is evolving, and new roles and opportunities are emerging. Here are a few examples:

Sustainability and green construction: Quantity surveyors can play a vital role in promoting sustainability and green construction practices. They can advise clients on sustainable design and construction options, prepare sustainable cost estimates, manage sustainable procurement, and monitor and report on sustainable performance.

Project management: Quantity surveyors have the skills and experience to be successful project managers. They can oversee all aspects of a construction project, from planning and budgeting to execution and completion.

Cost consulting: Quantity surveyors can provide cost consulting services to a variety of clients, including businesses, governments, and non-profit organizations. They can help clients to develop and manage budgets, estimate costs, and negotiate contracts.

Dispute resolution: Quantity surveyors are often called upon to resolve disputes between contractors and clients. They have the expertise to assess claims and make recommendations for resolution.

Risk management: Quantity surveyors can help clients to identify and manage risks associated with construction projects. They can develop risk management plans and identify strategies to mitigate risks.

Information technology: Quantity surveyors are increasingly using information technology to improve their efficiency and effectiveness. They are also using IT to develop new products and services for their clients.

Research and development: Quantity surveyors are involved in a variety of research and development projects. They are working to develop new tools and techniques for cost estimating, project management, and risk management.

Teaching and training: Quantity surveyors can share their knowledge and experience by teaching and training others. They can teach at universities and colleges, or they can provide training to professionals in the construction industry.

Quantity surveying in developing countries: Quantity surveyors are in high demand in developing countries. They can play a vital role in helping these countries to develop their infrastructure and economy.

Quantity surveying in the public sector: Quantity surveyors can work in the public sector, for governments and other public agencies. They can work on a variety of projects, such as roads, bridges, schools, and hospitals.

Quantity surveying in the private sector: Quantity surveyors can also work in the private sector, for construction companies, engineering firms, and other businesses. They can work on a variety of projects, from small residential projects to large commercial and industrial projects.

These are just a few examples of the emerging roles and opportunities for quantity surveyors. As the construction industry continues to evolve, new roles and opportunities will continue to emerge. Quantity surveyors who are adaptable and willing to learn new skills will be best positioned to take advantage of these new opportunities. The following are additional thoughts on the emerging roles and opportunities for quantity surveyors:

- Quantity surveyors can play a key role in the development and implementation of smart cities. Smart cities use technology to improve the efficiency and sustainability of urban living. Quantity surveyors can help to ensure that smart city projects are cost-effective and that they meet the needs of the community.

- Quantity surveyors can also play a role in the development of new technologies for the construction industry. For example, quantity surveyors can help to develop new software for cost estimating and project management. They can also help to develop new construction materials and methods.

- Quantity surveyors with a strong understanding of business can also pursue careers in management consulting. Management consultants help businesses to improve their performance and efficiency. Quantity surveyors can use their skills in cost estimation, project management, and risk management to help businesses to achieve their goals.

Overall, the future is bright for quantity surveyors. The profession is evolving and new roles and opportunities are emerging. Quantity surveyors who are adaptable and willing to learn new skills will be best positioned to take advantage of these new opportunities.

STEVEN SMITH PH.D.

CONCLUSION

The role of quantity surveyors is evolving rapidly, driven by a number of factors, including technological advancements, sustainability, and globalization. As a result, quantity surveyors are taking on new roles and responsibilities, such as becoming more involved in the early stages of construction projects, using technology to automate tasks, taking on a greater role in project management, becoming more involved in sustainability, and working on more international projects.

The evolving role of quantity surveyors presents a number of opportunities for the profession. Quantity surveyors who are adaptable and willing to learn new skills will be well-positioned to take advantage of these opportunities. The evolving role of quantity surveyors is a positive development. It is an opportunity for quantity surveyors to add more value to construction projects and to make a greater contribution to the industry. I am excited to see how the profession continues to evolve in the years to come.

The evolving role of quantity surveyors also presents some challenges. Quantity surveyors need to be willing to learn new skills and to adapt to new ways of working. They also need to be able to communicate effectively with a variety of stakeholders, including clients, architects, engineers, and contractors.

Overall, the future is bright for quantity surveyors. The profession is evolving and new roles and opportunities are emerging. Quantity surveyors who are adaptable and willing to learn new skills will be best positioned to take advantage of these new opportunities.

GLOSSARY
OF QUANTITY
SURVEYING TERMS

Bill of Quantities: A detailed document that lists and describes all the materials, quantities, and associated costs required for a construction project.

Cost estimate: A prediction of the total cost of a construction project.

Cost management: The process of planning, estimating, budgeting, and controlling costs throughout the lifecycle of a construction project.

Contract administration: The management of construction contracts, including the preparation, negotiation, and execution of contracts, as well as the administration of contracts during construction.

Dispute resolution: The process of resolving disputes between parties involved in a construction project.

Fee: The payment made to a quantity surveyor for their services.

Final account: A document that summarizes the final costs of a construction project.

Interim payment: A payment made to a contractor during the course of a construction project.

Project management: The process of planning, organizing, and controlling the resources necessary to complete a construction project on time, within budget, and to the required quality standards.

Quantity surveying: The profession of measuring and estimating the cost of construction work.

Retention: A sum of money held back from a contractor until the completion of a construction project.

Tender: An offer to undertake a construction project for a specified price.

Value engineering: A systematic approach to analyzing and optimizing the value of a project by balancing the cost, function, quality, and aesthetics of its components.

ADDITIONAL RESOURCES FOR FURTHER LEARNING

Professional organizations: Quantity surveying professional organizations offer a variety of resources for their members, including continuing education courses, publications, and networking opportunities. Some examples of quantity surveying professional organizations include:

- Royal Institution of Chartered Surveyors (RICS)
- Association for Cost Engineering International (AACEI)
- Australian Institute of Quantity Surveyors (AIQS)

Educational institutions: Many universities and colleges offer undergraduate and graduate programs in quantity surveying. These programs provide students with the knowledge and skills they need to become successful quantity surveyors.

Trade publications and journals: There are a number of trade publications and journals that cater to quantity surveyors. These publications provide readers with up-to-date information on the latest trends and developments in the quantity surveying profession. Some examples of quantity surveying trade publications and journals include:

- Construction Cost Engineering
- Cost Engineering
- Quantity Surveyor

Online resources: There are a number of online resources that provide information on quantity surveying. These resources include websites, blogs, and forums. Some examples of online quantity surveying resources include:

- Quantity Surveying Hub
- QS World
- Quantity Surveying Forum

ABOUT THE AUTHOR

Steven Smith, Ph.d.

Steven Smith is a renowned expert in the field of Construction Management, with a wealth of knowledge and experience spanning both academia and industry. Holding a doctorate in Construction Management, Steven has dedicated his career to advancing the field and contributing to its body of knowledge.

Throughout his academic journey, Steven's passion for understanding the intricacies of construction processes and finding innovative solutions to industry challenges became evident. His doctoral research focused on optimizing project management practices and enhancing productivity in construction projects, leading to a profound understanding of various aspects of construction management and their impact on project success.

BOOKS BY THIS AUTHOR

Construction Business Startup 101: Laying The Groundwork

Construction Business Startup 101

Are you ready to take the leap into construction entrepreneurship? This comprehensive guide will walk you through everything you need to know to start and run a successful construction business.

Inside, you'll discover:
How to craft a winning business plan that outlines your unique value proposition, target market, and financial projections.

How to build a strong team of skilled and experienced workers who are committed to your success.

How to manage finances effectively so you can stay on budget and profitable.

How to market and sell your services to attract new clients and projects.

How to overcome common challenges and setbacks that many construction businesses face.

You'll also learn from the experiences of successful construction entrepreneurs, so you can avoid their mistakes and learn from their successes. With this guide in your hands, you'll have everything you need to turn your vision into a reality and build a thriving construction business.

The Dictionary Of Construction Terminologies: A Compendium Of Knowledge For Students, Academics, Practitioners, And House Owners

The Dictionary of Construction Terminologies

Learn the language of construction from one of the most comprehensive dictionaries of construction terminologies available. From architecture and engineering to materials and equipment, this book covers several aspects of construction terminology in clear and concise language. With thousands of entries, the dictionary is an essential tool for anyone who wants to understand the complex world of construction.

The book features:

Comprehensive coverage of construction terminology

Clear and concise definitions, written in easy-to-understand language

Alphabetical organization for quick and easy reference

It is an essential tool for professionals, students, and anyone interested in the construction field.

www.ingramcontent.com/pod-product-compliance
Lightning Source LLC
Chambersburg PA
CBHW062355290526
45794CB00005B/2243